Gehad Hamdy
Ghalia A. Gaber

# Surface Engineering of Metal and Alloy

Gehad Hamdy
Ghalia A. Gaber

# Surface Engineering of Metal and Alloy

## Recent Implements of Metal and Alloy: Surface Engineering and Surface Modification Techniques

LAP LAMBERT Academic Publishing

**Imprint**



Cover image: www.ingimage.com

Publisher:
LAP LAMBERT Academic Publishing
is a trademark of
Dodo Books Indian Ocean Ltd. and OmniScriptum S.R.L publishing group

120 High Road, East Finchley, London, N2 9ED, United Kingdom
Str. Armeneasca 28/1, office 1, Chisinau MD-2012, Republic of Moldova, Europe
Managing Directors: Ieva Konstantinova, Victoria Ursu
info@omniscriptum.com

Printed at: see last page
**ISBN: 978-620-8-43909-5**

## Table of Contents

# 1. Introduction

## 1.1 Surface Engineering

A vast array of methods that seek to create and alter the surface characteristics of components are together referred to as surface engineering. To maximize the surface qualities and bulk materials, surface engineering techniques fall into two major groups. These are surface modifications and coatings. Surface form design is a component of other procedures. Applying a layer of molten, semi-molten, or chemical material to a substrate is the process of surface coating. Instead than changing the bulk material's composition, one of the primary purposes of surface coating is to enhance and alter the surface functionalities. Surface coating techniques include sol-gel, cladding, electroplating, plasma and thermal spraying, physical vapor deposition (PVD) [1], chemical vapor deposition (CVD) [2], and others [3]. Surface modification methods include diffusion treatments like carburizing, high-energy treatments like ion implantation, and hardening by flame, induction, laser, or electron beam. The components mechanical or physical characteristics can be altered, friction can be reduced and surface wear and corrosion resistance can be enhanced.

## 1.2 Surface and Engineering Environment

Because of the comparatively tiny fraction of load-supporting area, contact stresses rise when two surfaces come into contact and relative motion is created. Wear and friction will ensue from this, and failure may even follow. Certain parts of a modern engine or gearbox's high-stress applications allow heat to accumulate because the pressure of metal against metal pushes out all of the lubricating fluid. This heat is sufficient to instantly fuse the two pieces together in the event of high pressure and friction, right before their movement causes them to separate. The harmful effects of drag and wear are caused by this continuous weld and break process, which can happen at the atomic level. Thus, it is very advantageous to have surfaces that are engineered to combat friction and minimize wear.

Numerous materials have been created with special bulk qualities in mind, but their surface properties have not been given much attention. Using surface engineering, these issues can be resolved by:

i. Implanting alloying atoms to different depths, thereby improving toughness and fatigue properties; (surface modification)
ii. Depositing surface layers, thick or thin, including lubricants; (surface coating)
iii. Redesigning the surface shape of the component to distribute stresses.

## 1.3 Friction, Corrosion and Wear

Wear, corrosion, and friction issues are resolved using surface engineering approaches. The most frequent causes of engineering failures are wear, corrosion, and friction. However, a closer examination reveals that many tri-biological processes in geological and technological applications share characteristics with bio-system phenomena like surface passivation and the lotus leaf effect (hydrophobicity). Natural selection has resulted in surface textures and water-based lubricant systems that are significantly superior to the best oil-based lubricants used in the majority of fabricated devices [4]. The cost of friction, wear, and corrosion accounts for 7% of GNP in developed nations; effective tri-biological systems might potentially lower this amount by 1%. Globally, Europe uses 34% of all lubricants produced, of which 50% are recycled, while an additional 2.5 million tons are wasted in the environment.

Corrosion is mainly an electrochemical phenomenon, which occurs when metals react with materials in their environment. In order to create heterogeneous interfaces from hydrophilic materials, surface topographies that promote stable air/vapor entrapment must be carefully considered. When designing heterogeneous solid–liquid interfaces, two key factors are crucial. One is the surface's topology, while the other is the surface layer's composition. Roughness-induced heterogeneous wetting is a complex mechanism that affects different length scales. The stability of a composite interface is essential and should be addressed for the effective creation of corrosion-resistant surfaces since the transition to a homogenous contact is irreversible, making the composite interface delicate [5].

Wear is a surfaces gradual loss of substance. Abrasive and adhesive wear are more prevalent than other types of wear. Hard protuberances or particles that are pushed up against and move along a solid surface cause abrasive wear. Adhesive wear results from either material loss from one surface or localized bonding between contacting solid surfaces that transfers material between the two surfaces. One common way a soft material can erode a harder one is through adhesive wear. Adhesive and abrasive wear phenomena are illustrated in Figure 1.

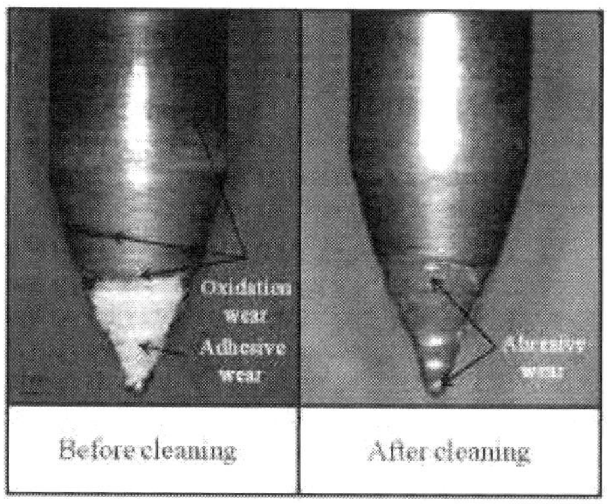


**Figure 1.** Demonstration of wear: Abrasive wear and Adhesive wear.

AlNiSi2 intermetallic is formed on the surface, increasing the samples strength and resistance to wear. The wetting effect of Cu samples has less of an impact on wear resistance than its creation during the sintering process. The results are shown visually in Figure 2. The findings reveal three primary phenomena. First, as the reinforcement ratio is gradually increased, the WR gradually drops; the lowest WR was observed for the 7.5 weight percent of the reinforcement particles. This showing that regardless of the coating type, adding more ceramic phase content enhances the composite material's overall wear resistance when compared to pure Al. As the Cu samples have good wetting and good distribution in the AMM, they must give a better wear resistance [6].

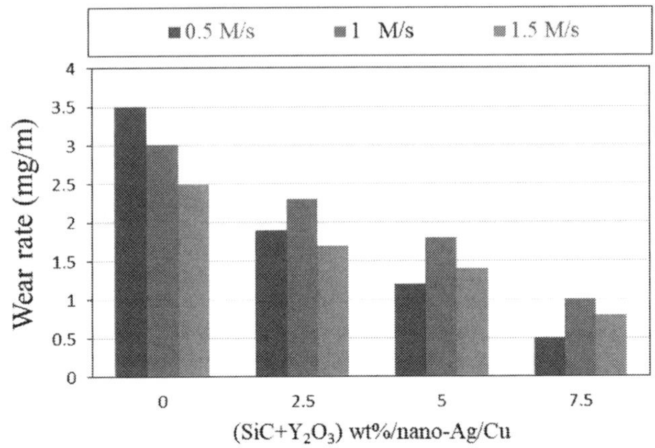


**Figure 2.** WR values of sintered Cu-coated samples with different reinforcement percentages [7].

### 1.4. Importance of Surface Modification and Morphology

The importance of surface morphology and surface modification in the field of materials science and engineering cannot be emphasized enough. Although materials bulk qualities frequently receive a lot of attention, their surface properties are crucial in determining their functioning, performance, and suitability for different applications. In order to highlight the significance of surface modification and morphology in the larger framework of interfacial engineering, this section attempts to clarify the critical role that these factors play in determining the characteristics and uses of materials [8].

Surface modification techniques encompass a diverse array of methodologies aimed at altering the chemical, physical, or morphological properties of material surfaces (Figure 3). These techniques enable engineers and scientists to tailor surface characteristics to meet specific requirements, such as enhanced adhesion, corrosion resistance, biocompatibility, or catalytic activity. Surface modification can involve chemical treatments, such as functionalization or coating deposition, physical techniques like plasma treatment or laser ablation, or biological approaches such as biomimetic coating or bio functionalization [9].

Electro-deposition of $GO/SiO_2$, $GO/TiO_2$, and $GO/ZrO_2$ nanocomposites in a 3.5% NaCl solution was used to determine the coating layer's anticorrosion behavior inhibitory effectiveness. In order to demonstrate the effectiveness of the enhanced oxide layer with $GO/ZrO_2$ as a self-cleaning coating with novel few-layer GO nanosheets with $GO/ZrO_2$, complementary electrochemical measurements such as weight reduction estimation, chronoamperometric current-time, and potentiodynamic polarization techniques were used [10].

The corrosion susceptibility of T91 steel frequently calls for the use of extremely effective coatings in a variety of industries. Although creating a stable, economical, and effective graphene covering for T91 steel has generated a lot of attention, it has been difficult to do. To solve this, Ni-reduced graphene doped with $Ni/GO/TiO_2$ and $GO/Fe\text{-}TiO_2$ nanocomposite coatings was applied to T91 steel substrates by electrodeposition. Crucially, the graphene nanocomposite coatings established robust Cr-C connections with the T91 steel surface, guaranteeing remarkable endurance. On T91 steel, the $Ni/GO/Fe\text{-}TiO_2$ coating demonstrated a corrosion rate (CR) of 0.854 mm/y at room temperature, which was significantly less than the CR of 2.321 mm/y for the untreated T91 steel [11].

The importance of surface modification lies in its ability to impart desirable properties to material surfaces, thereby expanding their utility and applicability across a wide range of domains. For instance, in the field of biomaterials, surface modification techniques are employed to render implantable devices biocompatible, reducing the risk of adverse reactions and improving tissue integration. Similarly, in the realm of electronics, surface modification can enhance the adhesion and conductivity of electrode materials, leading to improved performance and reliability in devices such as sensors and batteries [12].

Surface morphology refers to the topographical features and structures present on the surface of materials. These features can range from nanoscale roughness and textures to microscale patterns and architectures. Surface morphology plays a critical role in influencing various surface-dependent phenomena, including adhesion, wetting behavior, friction, and light scattering. By controlling surface morphology, researchers can manipulate these surface-dependent properties to achieve specific functionalities and performance characteristics [13]. The importance of surface morphology lies in its profound influence on material properties and behavior. For instance, in the field of optics, surface morphology dictates the reflectivity, scattering, and transmission properties of materials, making it crucial for applications such as antireflective coatings, photonic devices, and optical sensors. Similarly, in the domain of tribology, surface morphology influences friction and wear behavior, affecting the performance and longevity of mechanical components in diverse applications ranging from automotive engines to biomedical implants [14].

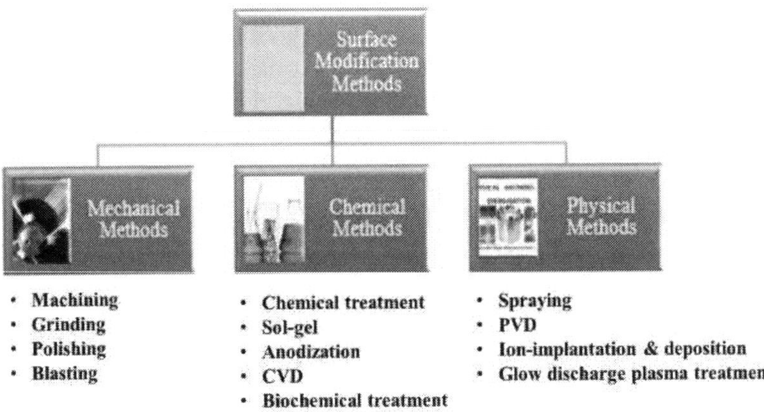


**Figure 3.** The available surface modification techniques.

In summary, the importance of surface modification and surface morphology in determining the applications and properties of materials cannot be overstated. These surface engineering strategies enable researchers to tailor material surfaces to meet specific requirements, thereby expanding the scope of material functionalities and enabling innovations across diverse fields. By leveraging surface modification and morphology within the broader framework of interfacial engineering, scientists and engineers can unlock new opportunities for advancing materials science and technology [14].

## 1.5. Classification of Nanomaterials and Their Properties

Nanomaterials represent a diverse class of materials characterized by their nanoscale dimensions, which impart unique and often enhanced properties compared to their bulk counterparts. These materials can be broadly classified based on their dimensionalities, compositions, and structural characteristics. One significant subclass of nanomaterials that has garnered considerable attention is nanofibers, distinguished by their high aspect ratios, large surface areas, and tunable properties [15].

Nanoparticles are one-dimensional nanomaterials with at least one dimension in the nanometer range. They encompass a wide variety of shapes and compositions, including spheres, rods, wires, and tubes. Nanoparticles exhibit size-dependent properties such as quantum confinement effects, surface plasm on resonance, and enhanced catalytic activity, making them valuable building blocks for diverse applications in electronics, catalysis, and biomedicine [16].

Nanocomposites are hybrid materials composed of a matrix phase and nanoscale reinforcement elements dispersed throughout. These reinforcements can take the form of nanoparticles, nanotubes, or nanofibers, and are incorporated into the matrix to impart enhanced mechanical, thermal, or electrical properties. Nanocomposites find applications in structural materials, coatings, and electronics, where their improved performance characteristics offer significant advantages over conventional materials [17][18].

Gaber et al. [19] use electrophoretic deposition (EPD) to deposit $SiO_2$ or $TiO_2$ nanoparticles in an anodic aluminum oxide layer (AAO) in order to study the electrochemical properties of aluminum alloy. The modified oxide layer exhibits promise for preventing corrosion while maintaining its ability to promote adhesion. The $TiO_2$-nanoparticle-modified oxide layer prevented corrosion better.

Nanowires and Nano rods are one-dimensional nanomaterials characterized by their elongated structures with diameters in the nanometer range and lengths extending to micrometers or even millimeters. These materials exhibit unique electrical, optical, and mechanical properties, with potential applications in Nano electronics, sensors, and energy conversion devices. By controlling the dimensions and compositions of nanowires and Nano rods, researchers can tailor their properties for specific applications [20].

Nanotubes are hollow cylindrical structures composed of carbon, boron nitride, or other materials, with diameters on the order of nanometers and lengths ranging from nanometers to millimeters. Carbon nanotubes, in particular, have garnered significant interest due to their exceptional mechanical strength, thermal conductivity, and electrical properties. They find applications in fields such as aerospace, electronics, and energy storage, where their lightweight and multifunctional characteristics offer unique advantages [21].

Nanofibers stand out as one-dimensional nanostructures characterized by their high aspect ratios, large surface areas, and nanoscale diameters. Fabricated from a variety of materials, including polymers, ceramics, and carbon-based materials, nanofibers are crafted using techniques such as electrospinning, template synthesis, and vapor-phase deposition. These remarkable nanostructures exhibit a plethora of unique properties that render them exceptionally versatile and valuable for a myriad of applications across diverse fields [18]. Nanofibers exhibit outstanding mechanical strength and flexibility, owing to their high aspect ratios and nanoscale dimensions. This inherent mechanical robustness makes them ideal candidates for reinforcement in composite materials, offering enhancements in structural integrity, impact resistance, and load-bearing capabilities.

Additionally, their large surface areas and porous structures make them highly effective in applications requiring high surface-to-volume ratios, such as filtration, catalysis, and sensing [23]. Furthermore, nanofibers possess exceptional electrical conductivity and thermal stability, stemming from their unique structural arrangement at the nanoscale. These properties render them invaluable for applications in electronics, energy storage, and thermal management, where their high surface-to-volume ratios and tunable conductive properties enable efficient charge transport and heat dissipation [24].

Beyond their mechanical, electrical, and thermal attributes, nanofibers exhibit remarkable biocompatibility and bioactivity, making them ideal candidates for applications in biomedicine and tissue engineering. Their nanoscale dimensions mimic the extracellular matrix found in biological tissues, facilitating cell adhesion, proliferation, and differentiation. Nanofiber-based scaffolds hold immense potential for regenerative medicine, wound healing, and drug delivery, offering tailored platforms for tissue regeneration and therapeutic intervention.

The versatility of nanofibers extends further to environmental applications, where their high surface areas and porous structures enable efficient adsorption and filtration of contaminants. Nanofiber-based materials have been employed for oil spill cleanup, air filtration, water purification, and environmental remediation, offering sustainable solutions to pressing environmental challenges.

Moreover, nanofibers find applications in food packaging, where their barrier properties and antimicrobial functionalities help extend the shelf life of perishable goods and ensure food safety. By incorporating nanofiber-based materials into packaging films and coatings, manufacturers can enhance the quality and integrity of food products, reducing waste and improving sustainability [25].

The classification of nanomaterials provides a framework for understanding their distinct properties and potential applications. Nanofibers, in particular, offer a wealth of opportunities for advancing interfacial engineering, owing to their unique structural characteristics and multifunctional properties. By exploring the interfaces within nanofiber-based materials, researchers can unlock new opportunities for tailoring their properties and functionalities, paving the way for innovative solutions in diverse fields.

# 2. Surface Modification Strategies

## 2.1. Chemical Functionalization

Chemical functionalization stands as a cornerstone in the realm of surface modification strategies, offering a versatile and precise approach to tailor material surfaces for specific applications. This strategy involves the introduction of functional groups onto the surface of a material, altering its chemical composition and reactivity. The process is achieved through a variety of chemical reactions, such as covalent bonding, that create robust and tailored interfaces [26]. One of the primary advantages of chemical functionalization lies in its ability to enhance the adhesion properties of a material. By introducing reactive groups that form strong bonds with other materials, surfaces become more amenable to adhesives, coatings, or other modifications. This is particularly crucial in industries where adhesion is a critical factor, such as in the development of advanced composites or biomedical devices.

Furthermore, chemical functionalization allows for the incorporation of specific functionalities onto the material surface. This is exploited in applications like biosensors, where the introduction of biomolecules or recognition sites onto the surface enables selective interactions with target. It also plays a pivotal role in improving the compatibility of materials in composite systems, enabling the development of materials with unique combinations of properties [27].

Table 1 outlines diverse surface modification strategies for nanomaterials. Chemical functionalization involves modifying surface chemistry, while physical coating alters surface characteristics. Nanoscale engineering tailors nanoscale features for specific functionalities, and self-assembly techniques create organized structures. These strategies find applications in nanocomposites, nanoscale devices, and drug delivery systems.

**Table 1.** Surface Modification Strategies in Nanomaterials

| Strategy | Description | Applications | Advantages | Limitations |
|---|---|---|---|---|
| Chemical functionalization | Covalent attachment of functional groups (molecules) to the surface of nanomaterials, altering their chemical and physical properties. | Nanoparticle functionalization for improved dispersion, biocompatibility, or targeting. - Development of functional nanocomposites with tailored properties. | High degree of control over surface chemistry. - Allows for specific functionalities to be introduced. | Can be complex and require specific reaction conditions. - May affect the inherent properties of the nanomaterial. |
| **Physical Coating and Deposition** | Deposition of thin films or coatings onto the surface of nanomaterials using various techniques (e.g., sputtering, evaporation). | Modifying surface properties like conductivity, corrosion resistance, or biocompatibility. - Creating functional devices like thin-film transistors or solar cells. | Versatile technique applicable to various materials. - Scalable for large-scale production. | Limited control over the thickness and uniformity of the coating. - May not be suitable for all types of nanomaterials. |
| **Nanoscale Engineering** | Tailoring the size, shape, and surface topology of nanomaterials at the nanoscale level to achieve desired functionalities. | Designing nanoscale devices with specific properties like sensors or catalysts. - Optimizing the performance of nanomaterials in various applications. | Allows for precise control over the functionality of nanomaterials. - Enables the creation of novel materials with unique properties. | Requires advanced fabrication techniques and expertise. - Can be expensive and time-consuming. |
| **Self-Assembly Techniques** | Spontaneous organization of | Assembly of nanomaterials | Bottom-up approach | Can be sensitive to |

| | | | | |
|---|---|---|---|---|
| | molecules into well-defined structures through non-covalent interactions (e.g., hydrogen bonding, van der Waals forces). | into complex structures for drug delivery systems. - Fabrication of functional materials with hierarchical structures. | offering high control over structure and functionality. - Can be used to create intricate and diverse structures. | environmental conditions and require careful control. - May not be suitable for large-scale production. |

## 2.2. Physical Coating and Deposition

Physical coating and deposition techniques offer a diverse set of tools to modify material surfaces, relying on the application of thin films or coatings to alter surface properties. Techniques such as physical vapor deposition (PVD) and chemical vapor deposition (CVD) enable the controlled deposition of materials onto a substrate, resulting in thin films that impart specific functionalities [28]. Physical coating is widely utilized to enhance the mechanical, optical, and corrosion-resistant properties of surfaces. For instance, the deposition of protective coatings on metal surfaces can mitigate corrosion, extending the material's lifespan. In optics, thin films can be engineered to control light reflection and transmission, contributing to the development of anti-reflective coatings or reflective mirrors [29][30]. Deposition techniques are also instrumental in the semiconductor industry for the fabrication of thin film transistors, solar cells, and other electronic devices. By precisely controlling the thickness and composition of the deposited films, engineers can optimize the electrical and optical properties of the materials, enabling the development of high-performance electronic components.

## 2.3. Nanoscale Engineering

Nanoscale engineering represents a frontier in surface modification, where the manipulation of materials at the atomic and molecular levels opens up unprecedented possibilities for tailoring material interfaces. This strategy leverages nanotechnology to engineer surfaces with specific nanoscale features, influencing properties such as conductivity, catalytic activity, and mechanical strength. Nanoscale engineering is particularly influential in the development of nanocomposites, where Nano-sized fillers, such as nanoparticles or nanotubes, are incorporated into a matrix material. The high surface area and unique properties of these nanoscale fillers impart exceptional strength, conductivity, or other desired characteristics to the composite material. This strategy has applications in industries ranging from aerospace, where lightweight and high-strength materials are crucial, to electronics, where nanocomposites contribute to the development of advanced electronic devices.

Moreover, nanoscale engineering plays a pivotal role in the fabrication of Nano devices and sensors. By precisely controlling the size, shape, and composition of nanostructures, researchers can create materials with tailored electronic, optical, or sensing properties. This has implications for the development of efficient sensors, advanced imaging technologies, and other cutting-edge applications [31].

## 2.4. Self-Assembly Techniques

Self-assembly techniques harness the inherent properties of materials to spontaneously organize into well-defined structures or patterns. This strategy takes inspiration from nature, where molecular and supramolecular self-assembly processes play a fundamental role in the formation of complex biological structures. By leveraging self-assembly, researchers can engineer materials with precise control over their structure and functionality. One of the key advantages of self-assembly lies in its ability to create intricate patterns and structures at multiple length scales. This is particularly relevant in the development of nanomaterials and Nano devices, where precise control over the arrangement of components is essential. Self-assembly techniques are employed in the fabrication of nanoscale patterns for electronics, sensors, and other applications [32]. Furthermore, self-assembly strategies are integral in the development of biomimetic materials. By mimicking natural processes, such as the self-assembly of proteins or lipids, researchers can create materials with tailored functionalities. This has applications in drug delivery systems, where self-assembled structures can enhance drug stability and targeting.

In terms of green chemistry, the use of the use of expired drugs has recently gained attention as sustainable corrosion inhibitors due to their biodegradability, environmental friendliness, low cost, and easy availability. Recycling expired drugs as corrosion inhibitors reduce economic losses and waste management costs. Expired or unused drugs harm humans, aquatic creatures, and terrestrial ecosystems [33]. Research reveals that expired medicine prevented metal corrosion by acting as a barrier. Drugs that suppress corrosion are safe, cheap, and eco-friendly.

Gaber et al. [34] evaluate corrosion rate (CR) and inhibitor efficiency (IE) of Cu–Zn alloy in SCW with a high concentration of Bepotastine-B and found to be (0.625m/y) and (94.44%), respectively. According to the findings of previous research, expired pharmaceuticals are potential materials that can be employed as green corrosion inhibitors. Overall, research investigations show that using pharmaceuticals as corrosion inhibitors is not only the best option for preserving metals from corrosion, but it also leads to waste management of expired drugs.

In conclusion, surface modification strategies are paramount in tailoring material interfaces for diverse applications. Chemical functionalization provides a versatile approach for introducing specific functionalities, physical coating and deposition enable controlled film formation, nanoscale engineering opens avenues for manipulating materials at the nanoscale, and self-assembly techniques offer a unique way to create ordered structures. The integration of these strategies empowers scientists and engineers to design advanced materials with tailored properties, driving innovation across industries and contributing to the development of next-generation technologies.

# 3. Applications of surface engineering

Surface engineering plays a crucial role in various industrial sectors, enabling the enhancement of material performance, protection against environmental degradation, and the development of advanced functionalities. This section focuses on the applications of surface engineering in key industries, including aerospace and defense, automotive, electronics and semiconductor, biomedical and healthcare, and energy sectors. It discusses the specific applications, challenges, and benefits of surface engineering techniques in each industry.

## 3.1. Surface Engineering in Polymer Materials

Polymer blends and composites represent a fascinating arena within the realm of interfacial engineering, where the combination of different polymers or the integration of polymers with other materials leads to the creation of materials with tailored properties. The blending or compositing of polymers provides a versatile approach to achieve synergistic effects, combining the strengths of individual components while mitigating their weaknesses. In polymer blends, two or more polymers are physically mixed to form a single material. The interfaces between the different polymer phases become crucial in determining the overall performance of the blend. Interfacial engineering strategies are employed to enhance compatibility between the polymers, improving the blend's mechanical, thermal, and morphological properties.

For instance, the addition of compatibilizing agents or the modification of polymer surfaces through chemical functionalization can enhance the adhesion between polymer phases, reducing phase separation and improving overall material properties [35]. Polymer composites, on the other hand, involve the incorporation of reinforcing materials, such as fibers, nanoparticles, or other polymers, into a polymer matrix. The interfaces between the matrix and the reinforcing materials play a pivotal role in determining the composite's strength, stiffness, and other mechanical properties.

Polymer/composite materials have been commonly used as corrosion inhibitors in different fields such as marine, oil field, and engineering industries due to their self-healing and thermal stability features [36]. Materials such as recycled plastics, metals, and paper reduce the need for virgin resources and help divert waste from landfills. Using eco-friendly materials in various industries, such as packaging, construction, and biomedical applications. Sustainable design and manufacturing practices prioritizing using eco-friendly materials contribute to a more circular economy by promoting resource efficiency and waste reduction [37].

Dardeer et al. [38] developing a novel chitosan-Schiff base and its derivatives is a distinct and appealing synthetic route for green corrosion inhibitors to lower the rate of Cu-Ni alloy industrial corrosion. Cu-Ni alloy is a material of choice for condensers and heat exchangers, where seawater is used as a coolant and in desalination plants [39]. Therefore, a novel chitosan-sodium pyruvate (Chi-SP) polymer was successfully prepared through the condensation reaction of chitosan with sodium pyruvate. (Chi-SP) polymer is tailored into the hydrophobic core of β-cyclodextrin (CD) by two routes (chemically and physically) to give a novel Chi-SP/β-CD inclusion complex (IC), Chi-SP/β-CD physical mixture (PM), respectively.

The physicochemical properties and the corrosion inhibition efficiency (CIE) of Chi-SP, Chi-SP/β-CDIC, Chi-SP/β-CD PM, and Chi-SP/$TiO_2$ were performed. The corrosion rate (CR) and inhibitor efficiency (IE) of Cu-Ni alloy in HCl (1 M) using Chi-SP, PM, IC, and the nanocomposite were found to be (82.53, 61.63, 65.56, and 62.94) % and (54.1, 65.7, 63.5, and 65) %, respectively. These findings reveal that the Chi-SP/$TiO_2$ molecules have many polar hydroxyl groups, which can contribute to the chemisorption and physisorption of the inhibitor on the alloy surface.

Surface modification of polymers is a key facet of interfacial engineering that focuses on altering the properties of the polymer surface to meet specific requirements. The surface characteristics of polymers strongly influence their interactions with other materials, adhesion properties, and overall performance in various applications. By tailoring the polymer surface, scientists and engineers can enhance functionalities such as wettability, adhesion, and biocompatibility. One prevalent method of surface modification is chemical functionalization, where specific functional groups are introduced onto the polymer surface. This can be achieved through techniques like plasma treatment, grafting, or coating. For example, introducing hydrophilic groups can improve the wettability of a polymer surface, making it more receptive to coatings or adhesives. Conversely, hydrophobic functionalization can be applied to repel water, a desirable trait in applications such as water-resistant coatings or biomedical implants [40].

Physical methods, such as laser ablation or ion beam irradiation, are also employed for surface modification. These techniques alter the surface topography and chemistry, affecting properties such as roughness, adhesion, and optical characteristics. Surface modifications are extensively utilized in industries ranging from packaging, where improved adhesion properties.

### 3.2. Surface Engineering in Nanomaterials

Surface engineering in nanomaterials stands at the forefront of scientific innovation, driving the development of materials with unprecedented properties and functionalities. At the nanoscale, the surfaces and interfaces of nanoparticles play a pivotal role in dictating their behavior, interactions, and applications. Interfacial engineering strategies in this domain aim to harness and manipulate these surfaces to design nanomaterials with tailored properties. Nanoparticles, with dimensions typically ranging from 1 to 100 nanometers, exhibit unique properties compared to their bulk counterparts. The high surface-to-volume ratio of nanoparticles makes their surfaces particularly reactive and sensitive to environmental conditions.

Interfacial engineering focuses on optimizing this surface reactivity for specific applications. For example, functionalizing the surface of nanoparticles with ligands or functional groups enhances their stability, and compatibility with various solvents or matrices. Furthermore, the interfaces between nanoparticles within a system, often referred to as interparticle interfaces, are crucial in determining the overall behavior of nanomaterials. Interparticle interactions influence phenomena such as aggregation, self-assembly, and the formation of nanocomposites. Tailoring these interfaces through surface modifications or coatings allows for the precise control of nanoparticle dispersion and alignment, critical in achieving desired material properties [41].

Nanocomposites and hybrid materials represent a cutting-edge application of interfacial engineering in nanomaterials, where nanoparticles are integrated into a matrix material to create a material with synergistic properties. The interfaces between the nanoparticles and the matrix play a decisive role in governing the overall performance of these nanocomposites. In nanocomposites, nanoparticles are dispersed within a polymeric, metallic, or ceramic matrix. The interfacial interactions between the nanoparticles and the matrix are crucial for achieving desirable mechanical, thermal, and electrical properties. Effective interfacial engineering involves enhancing the compatibility between the nanoparticles and the matrix, ensuring a strong interfacial adhesion. This is often achieved through surface functionalization of the nanoparticles or the use of coupling agents that promote a robust interface between the two components.

Hybrid materials, on the other hand, involve the integration of different types of nanomaterials into a single system. The interfaces between these different nanomaterials contribute to the synergistic effects observed in hybrids. For instance, combining metallic nanoparticles with carbon nanotubes in a hybrid material can lead to enhanced conductivity and mechanical strength compared to individual components. Interfacial engineering strategies focus on optimizing the compatibility and interactions between these diverse nanomaterials to achieve the desired combination of properties [42][43].

Surface engineering at the nanoscale transcends mere physical boundaries, encompassing the intricate interactions that occur within materials at this level. These nanoscale interactions exert a profound influence on material behavior and pave the way for groundbreaking applications across various fields. Understanding and manipulating these interactions offer unprecedented opportunities for innovation and advancement [44].

In healthcare, nanomaterials engineered at the interface of biology and nanotechnology have revolutionized diagnostics and drug delivery. Nanoparticles coated with biocompatible materials, such as lipids or polymers, navigate biological barriers more effectively, enabling targeted drug delivery. The surface modifications of these nanoparticles play a crucial role in ensuring stability, biocompatibility, and controlled release of therapeutic agents. Moreover, surface-engineered nanomaterials offer exciting possibilities for imaging, bio sensing, and personalized medicine, heralding a new era in medical diagnostics and treatment modalities [45].

Nanocatalysts, with their high surface area and unique reactivity, find extensive use in catalysis. Surface engineering in this context involves optimizing the surface composition and structure of nanoparticles to enhance catalytic activity. Tailoring the interfaces between the catalytic nanoparticles and the reaction environment improves efficiency and selectivity in various chemical processes. These advancements in nanocatalysis hold promise for sustainable energy production, environmental remediation, and the synthesis of high-value chemicals [46].

In the realm of energy storage, nanomaterials offer significant potential for advancing battery and capacitor technologies. The interfaces between nanomaterials and electrode surfaces play a critical role in charge storage and transport. By engineering these interfaces, researchers aim to enhance the energy density, cycling stability, and overall performance of energy storage devices. Nanomaterial-based electrodes, electrolytes, and separators hold promise for next-generation batteries and supercapacitors, addressing the growing demand for efficient energy storage solutions in portable electronics, electric vehicles, and renewable energy systems [47].

Nanomaterials offer innovative solutions for environmental challenges, such as pollution and water purification. Functionalized nanoparticles can selectively capture pollutants or catalyze degradation reactions. Surface engineering in these applications focuses on optimizing the surface chemistry to enhance pollutant adsorption and promote catalytic activity. Nanomaterial-based sensors and filtration systems hold promise for real-time monitoring of environmental pollutants and the remediation of contaminated air and water sources, contributing to global efforts towards sustainability and environmental stewardship [48][49].

Nanomaterials are pivotal in the development of advanced electronics and optoelectronics. Quantum dots, for instance, exhibit unique electronic and optical properties at the nanoscale. Surface engineering in these materials involves controlling the size, shape, and composition of quantum dots to tailor their electronic band structure and light-emitting properties. This has significant implications for the development of high-performance displays, sensors, and solar cells, enabling advances in information technology, telecommunications, and renewable energy generation [50].

In conclusion, surface engineering in nanomaterials represents a dynamic and multidisciplinary field with immense potential for scientific and technological advancements. Whether manipulating the surfaces of nanoparticles for improved dispersion and reactivity, engineering interfaces in nanocomposites to achieve enhanced material properties, or harnessing nanoscale interactions for innovative applications, the strategic control of interfaces at the nanoscale opens up new frontiers in materials science and engineering. As research in this field continues to progress, we anticipate an even greater impact on diverse industries and the development of transformative technologies.

### 3.3. Interfacial Engineering for Electronic Devices

In the realm of electronic devices, semiconductor interfaces lie at the heart of their functionality. Surface engineering plays a critical role in optimizing these interfaces to enhance the performance, reliability, and efficiency of semiconductor-based devices. Semiconductors are the building blocks of electronic devices, and their interfaces with various materials, including insulators, metals, and other semiconductors, significantly influence device behavior. One of the primary applications of surface engineering in semiconductor devices is the modulation of charge carrier transport across semiconductor-insulator interfaces. In field-effect transistors (FETs), for instance, the semiconductor-insulator interface plays a pivotal role in controlling the flow of electrons or holes. Interfacial engineering strategies focus on reducing interface states and defects, improving dielectric properties, and enhancing charge injection and extraction. This optimization is crucial for achieving higher carrier mobility, lower leakage currents, and improved switching characteristics in transistors [51].

Moreover, semiconductor interfaces are fundamental in the development of diodes, where the junction between different semiconductor materials allows for controlled rectification of current flow. Surface engineering in diode devices involves precise control over the semiconductor-dielectric interfaces, ensuring efficient charge carrier injection and extraction. This optimization is essential for achieving high rectification ratios, low reverse leakage currents, and stable diode characteristics. Semiconductor interfaces are optimized for improved charge transport, while surface engineering in photovoltaics enhances energy conversion. Applications in flexible electronics involve tailoring interfaces for flexibility and deformation resistance, contributing to advancements in wearable electronics and flexible displays.

The quest for efficient and sustainable energy sources has propelled the field of photovoltaics to the forefront of technological innovation. Surface engineering plays a pivotal role in optimizing the interfaces within photovoltaic devices, such as solar cells, to maximize energy conversion efficiency. Photovoltaic devices primarily rely on the generation and separation of charge carriers at semiconductor, making surface engineering paramount for their performance. In solar cells, the interface between the semiconductor and the photoactive layer, often composed of organic or inorganic materials, is critical for efficient light absorption and charge carrier extraction. Surface engineering strategies involve surface modifications, such as passivation layers or doping profiles, to enhance charge carrier mobility and reduce recombination losses.

This optimization is crucial for achieving higher power conversion efficiencies in solar cells. Moreover, the interfaces within multifunction solar cells, composed of different semiconductor materials with varying bandgaps, require precise engineering for optimal energy harvesting across the solar spectrum. Surface engineering in this context involves the careful design of semiconductor layers to facilitate efficient charge transfer and minimize losses at the interfaces between different materials. This is particularly relevant in advanced solar cell technologies, including tandem and perovskite solar cells, where multiple interfaces contribute to overall device performance [52].

The emergence of flexible electronics has revolutionized the design and fabrication of electronic devices, paving the way for innovative applications such as flexible displays, wearable sensors, and conformable electronic circuits. Surface engineering is instrumental in realizing the full potential of flexible electronics by optimizing the interfaces within flexible devices to withstand mechanical deformation while maintaining electronic performance. One key aspect of surface engineering in flexible electronics involves the development of flexible and stretchable substrates, often composed of polymers or elastomers. These substrates form the foundation for flexible devices and host the interfaces between various electronic components. Surface engineering strategies focus on enhancing the adhesion between flexible substrates and functional layers, such as semiconductors, conductors, and dielectrics, to ensure mechanical robustness and reliable electronic performance under deformation. In addition to substrate interfaces, surface and interfacial engineering is crucial in the design of flexible interconnects and electrodes.

The interfaces between conductive materials and flexible substrates must accommodate mechanical strain without compromising electrical conductivity.

Strategies such as nanomaterial-based inks, compliant electrode designs, and engineered interfaces contribute to the development of flexible and stretchable conductors. Furthermore, in flexible display technologies, surface and interfacial engineering is employed to optimize the interfaces within organic light-emitting diodes (OLEDs) and thin-film transistors (TFTs). Flexible OLED displays require engineered interfaces to ensure efficient charge injection, transport, and recombination, while flexible TFTs demand optimized semiconductor-dielectric interfaces to enable reliable transistor operation under mechanical strain. Surface engineering in flexible electronics extends beyond material interfaces to include the design of flexible encapsulation layers. These layers protect electronic components from environmental factors such as moisture and oxygen, ensuring the long-term reliability of flexible devices. Strategies such as barrier coatings and conformal encapsulation contribute to the development of flexible electronics with enhanced durability and stability [53].

In summary, surface engineering plays a transformative role in the development of electronic devices, ranging from semiconductor interfaces crucial for transistor operation to optimized interfaces in photovoltaic devices for efficient energy conversion, and flexible electronics where interfaces between materials enable deformation without compromising performance. As electronic devices continue to evolve, the strategic control and engineering of interfaces will remain at the forefront of innovation, driving advancements in electronic technology for diverse applications.

### 3.4. Biomedical Applications of Surface Engineering

The intersection of materials science and biology has given rise to a burgeoning field with immense potential: biomedical applications of interfacial engineering. This multidisciplinary approach focuses on tailoring the interfaces of materials used in biomedical contexts to enhance their performance, biocompatibility, and interactions with biological systems. In this exploration, we delve into key aspects of biomedical interfacial engineering, spanning surface modification in biomaterials, drug delivery systems, and interface design for biocompatibility [54]. Surface modification strategies enhance biocompatibility, drug delivery systems tailor interfaces for controlled drug release, and interface design minimizes immunogenicity for improved tissue integration. These aspects find applications in implants, tissue engineering, and neural interfaces.

Biomaterials, ranging from implants to scaffolds, play a crucial role in modern medicine, and their surface properties significantly influence their interactions with biological tissues. Interfacial engineering strategies aim to optimize these surfaces for improved biocompatibility, reduced immunogenicity, and enhanced functionality [55]. Figure 4 illustrates various surface modification strategies employed in biomaterials for enhanced biocompatibility and functionality. The central diagram displays techniques such as chemical functionalization, biomimetic coatings, and antimicrobial surface treatments. Insets highlight specific examples of each strategy and their impact on surface properties.

Surface modification is a critical aspect of biomaterial design, influencing interactions with biological systems. Chemical functionalization enables tailored surface chemistries, promoting cell adhesion and tissue integration. Biomimetic coatings replicate natural environments, fostering bioactivity and tissue regeneration. Antimicrobial treatments mitigate infection risks in implanted devices. This figure emphasizes the diversity of surface modification strategies for biomaterials.

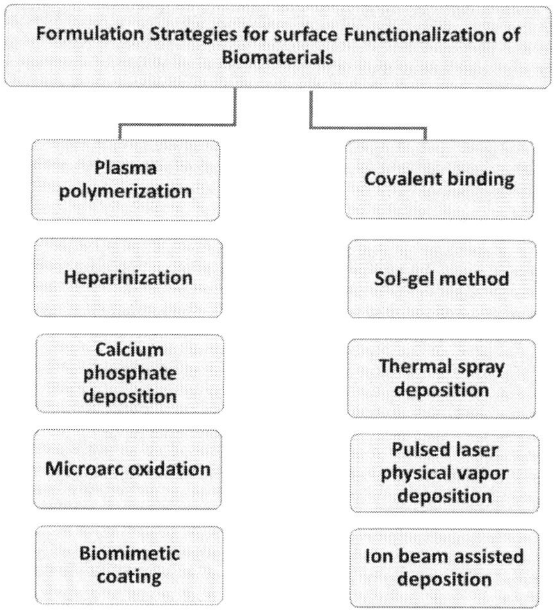


**Figure 4.** Surface Modification Strategies in Biomaterials.

Surface modification in biomaterials often begins with the alteration of surface chemistry to create interfaces that encourage favorable cellular responses. Techniques such as plasma treatment, chemical grafting, or coating with biomimetic molecules allow for precise control over surface functionalities. Hydrophilic coatings, for instance, can improve the wettability of implants, promoting better cell adhesion and tissue integration [56].

Biomimetic coatings aim to replicate the natural environment of tissues, fostering enhanced interactions between the biomaterial and the biological system. This includes the incorporation of bioactive molecules, such as peptides or growth factors, onto the biomaterial surface to mimic the extracellular matrix and guide cellular behavior. These coatings are instrumental in promoting tissue regeneration and minimizing adverse reactions [57].

In the context of implantable devices, preventing microbial infections is a critical consideration. Interfacial engineering can be employed to create antimicrobial surfaces that inhibit bacterial adhesion and biofilm formation. Functionalizing biomaterial surfaces with antimicrobial agents, such as silver nanoparticles or antimicrobial peptides, provides an effective strategy to mitigate the risk of infections associated with implants [58].

The realm of drug delivery systems has undergone a transformative shift with the advent of interfacial engineering. By tailoring interfaces between drug carriers and biological systems, researchers can achieve precise control over drug release kinetics, targeting capabilities, and therapeutic efficacy [59].

Nanoparticles, such as liposomes, micelles, and polymeric nanoparticles, serve as versatile drug carriers with tunable interfaces. Interfacial engineering in nanoparticle-based drug delivery involves optimizing surface properties to enhance biocompatibility and achieve targeted drug delivery. Surface modifications can include the attachment of targeting ligands or the addition of stealth coatings, such as polyethylene glycol (PEG), to evade the immune system and prolong circulation times.

Hydrogels and scaffolds provide three-dimensional matrices for drug delivery and tissue engineering applications. Interfacial engineering strategies focus on tailoring the surface properties of these materials to optimize drug loading, release kinetics, and interactions with surrounding tissues. Functionalizing hydrogels with bioactive molecules or modifying scaffold surfaces with specific ligands enhances their capacity to support cellular adhesion, proliferation, and differentiation.

Responsive drug delivery systems leverage interfacial engineering to create stimuli-responsive materials that release drugs in response to specific triggers. These triggers can include changes in pH, temperature, or the presence of specific biomolecules. Such systems offer precise control over drug release, minimizing side effects and improving therapeutic outcomes. Interfacial engineering is crucial in designing responsive interfaces that can sense and respond to these external stimuli.

Biocompatibility, the ability of a material to interact with biological systems without causing harm, is a fundamental consideration in biomedical applications. Interfacial engineering strategies focus on designing interfaces that promote favorable interactions with living tissues and minimize adverse reactions [60].

In the context of implants and medical devices, minimizing immunogenicity is paramount. Interfacial engineering can be employed to create biocompatible interfaces that reduce the activation of the immune system. Strategies include the use of anti-fouling coatings, such as zwitterion materials, that resist protein adsorption and cell adhesion, thereby minimizing inflammatory responses.

In tissue engineering, where the goal is to regenerate or replace damaged tissues, interfacial engineering plays a pivotal role in designing interfaces that support cellular integration and tissue formation. Biomaterial scaffolds are engineered with interfaces that mimic the natural extracellular matrix, providing cues for cell adhesion, migration, and differentiation. This facilitates the regeneration of functional tissues.

Neural surface, such as those used in neuroprosthetics or brain-machine interfaces, require precise interfacial engineering to ensure seamless communication between the device and neural tissues. Electrodes with optimized interfaces facilitate reliable signal recording and stimulation. Materials with enhanced biocompatibility and tailored surface properties are crucial to minimize tissue response and ensure long-term functionality [61]. While significant strides have been made in the field of biomedical interfacial engineering, challenges persist.

The biocompatibility of materials over extended periods, the potential for immune responses, and the need for long-term stability are among the ongoing considerations. Future directions in this field include the development of smart interfaces that dynamically respond to the biological environment, the incorporation of advanced materials with enhanced functionalities, and the integration of nanotechnology for precise control over interfaces.

In conclusion, the biomedical applications of interfacial engineering represent a frontier of innovation with transformative implications for healthcare. Whether optimizing surfaces of biomaterials, enhancing drug delivery systems, or designing interfaces for biocompatibility, the strategic control of interfaces plays a pivotal role in advancing medical treatments and technologies. As researchers continue to unravel the complexities of biological interactions at material interfaces, we anticipate further breakthroughs that will shape the future of biomedical engineering and contribute to improved patient outcomes.

## 4. Challenges and Future Directions

Surface engineering has made significant strides in enhancing the performance and functionality of metals in various applications. However, there are still challenges to be addressed and opportunities for future advancements in the field. This section discusses the limitations of current surface engineering techniques, emerging trends and technologies, and potential areas for future research and development.

### 4.1. Limitations of Current Surface Engineering Techniques

Many surface engineering techniques involve complex processes, specialized equipment, and skilled operators, resulting in high production costs. The implementation of these techniques in industrial settings may be challenging due to the need for specialized facilities and expertise. There is a need to develop cost-effective and scalable surface engineering processes that can be readily adopted by industries.

While surface coatings and modifications can enhance the performance of metals, their durability and longevity can still be a concern. Coatings may degrade over time, leading to reduced protection against wear, corrosion, or other environmental factors. Improving the longevity of surface-engineered materials is crucial to ensure long-term performance and cost-effectiveness.

Achieving good adhesion between the coating and the substrate is essential for the effectiveness of surface engineering techniques. However, certain combinations of materials may pose challenges to achieving strong and durable adhesion. Developing strategies to improve the compatibility and adhesion between coatings and substrates is crucial for reliable and long-lasting surface engineering solutions.

Surface engineering techniques may face limitations when it comes to treating complex geometries or intricate structures. Coating uniformity and coverage can be difficult to achieve in such cases, leading to variations in performance across the surface. Finding effective solutions for the surface engineering of complex shapes and structures is necessary to meet the demands of diverse industries and applications.

Some surface engineering processes involve the use of hazardous chemicals or high-energy inputs or generate waste products, which can have negative environmental implications. Minimizing the environmental footprint of surface engineering techniques is crucial to ensure sustainable and responsible manufacturing practices. Developing eco-friendly alternatives and optimizing process efficiency are important considerations.

Certain surface engineering techniques may have limitations in terms of the materials they can be applied to. For example, some coatings may not adhere well to certain types of metals or non-metallic materials. Expanding the range of materials that can benefit from surface engineering techniques is essential to cater to a broader range of applications and industries.

Achieving consistent and reproducible results in surface engineering can be challenging, particularly when it comes to large-scale production. Variations in process parameters, environmental conditions, or material properties can affect the quality and performance of surface-engineered products. Enhancing process control and developing robust quality-assurance protocols are necessary to ensure reliability and reproducibility in surface engineering.

Incorporating surface engineering techniques into existing manufacturing workflows can be complex and may require modifications to production lines or equipment. The seamless integration of surface engineering processes with existing manufacturing systems is crucial to facilitate their widespread adoption and enable efficient production. Addressing these limitations will require ongoing research and development efforts in the field of surface engineering. Collaborations between researchers, engineers, and industry professionals are essential to overcome these challenges and drive the advancement of surface engineering techniques towards greater efficiency, affordability, durability, and sustainability. A list of challenges and future directions in surface engineering of metals is presented in Table 2.

**Table 2.** Challenges and future directions in surface engineering of metals.

| Challenges | Future directions |
|---|---|
| **Process Complexity and Cost** | – Develop cost-effective and scalable surface engineering processes<br>– Simplify process steps and equipment requirements<br>– Improve process efficiency and productivity |
| **Durability and Longevity** | – Enhance coating durability and longevity<br>– Develop self-healing or protective coatings<br>– Optimize coating performance under various environmental conditions |
| **Compatibility and Adhesion** | – Improve compatibility between coatings and substrates<br>– Develop advanced surface preparation techniques for better adhesion<br>– Enhance interfacial bonding and mechanical properties |
| **Complex Geometries** | – Develop specialized surface engineering techniques for complex shapes and structures<br>– Explore additive manufacturing for surface engineering of complex geometries<br>– Optimize coating deposition methods for uniform coverage |
| **Environmental Impact** | – Minimize the use of hazardous chemicals and reduce energy inputs<br>– Develop eco-friendly alternatives and sustainable manufacturing processes<br>– Implement waste reduction and recycling methods for surface-engineered materials |
| **Limited Material Selection** | – Expand the range of materials suitable for surface engineering<br>– Develop coatings compatible with non-metallic materials<br>– Explore surface engineering techniques for polymers and composites |
| **Process Control and Reproducibility** | – Enhance process control and optimization<br>– Develop quality-assurance protocols and standards<br>– Implement real-time monitoring and feedback systems |
| **Integration with Existing Manufacturing Processes** | – Facilitate seamless integration of surface engineering into existing production workflows<br>– Modify production lines or equipment to accommodate surface engineering processes<br>– Develop standardized protocols for integrating surface engineering techniques |

### 4.2. Emerging Trends and Technologies in the Field

There is a growing interest in the development of multifunctional coatings that offer a combination of properties, such as self-healing, self-cleaning, anti-icing, or antifouling capabilities. These coatings can provide enhanced performance and reduce the need for additional treatments or maintenance. The integration of different functionalities into a single coating represents an emerging trend in surface engineering [62].

Nanotechnology has revolutionized surface engineering by enabling precise control over surface structures and properties at the nanoscale. The use of nanomaterials, such as nanoparticles, nanocomposites, and Nano layers [63], has shown promise in improving the performance of surface-engineered materials. The incorporation of nanotechnology into surface engineering opens up new opportunities for tailoring properties, enhancing functionality, and developing advanced applications.

Inspired by nature, biomimetic surfaces aim to replicate the unique properties and functionalities found in natural systems. The development of surfaces that mimic the self-cleaning properties of lotus leaves, the adhesive properties of gecko feet, or the drag-reduction properties of sharkskin represents an exciting direction in surface engineering. Biomimetic surfaces have the potential to provide innovative solutions in fields such as aerospace, automotive, and healthcare [64].

Additive manufacturing, or 3D printing, is being increasingly explored as a tool for surface engineering. This technology allows for the precise deposition of materials layer by layer, enabling the creation of complex surface structures and coatings with tailored properties. Additive manufacturing techniques can be used to fabricate surfaces with graded compositions, roughness profiles, or even embedded functionalities, offering new possibilities for customized surface engineering solutions [65].

The development of smart surfaces is an emerging trend in surface engineering. These surfaces have the ability to actively respond and adapt to changes in their environment. For example, smart surfaces can change their wettability, adhesion, or optical properties in response to external stimuli such as temperature, light, or pH. Such surfaces hold promise for applications in fields such as adaptive optics, microfluidics, and sensors, where dynamic control over surface properties is crucial [66].

The integration of machine learning and artificial intelligence (AI) techniques into surface engineering is gaining attention. These technologies can assist in the design and optimization of surface structures, coatings, and modifications by predicting their properties and performance based on large datasets and computational modeling. Machine-learning algorithms can also aid in the development of new materials with tailored surface properties, enabling faster and more efficient surface engineering processes [67].

Real-time characterization and monitoring of surface properties and performance are becoming increasingly important in surface engineering. In situ techniques such as spectroscopy, microscopy, and sensing methods allow for the assessment of surface modifications and coatings during operation or in simulated environments. This information can guide the optimization of surface engineering processes, ensure the quality and reliability of surface treatments, and enable proactive maintenance strategies [68].

### 4.3. Potential Areas for Future Research and Development

With the increasing focus on sustainability, there is a need for surface engineering techniques that minimize the environmental impact. Research efforts should be directed towards the development of environmentally friendly coatings, the reduction of hazardous materials, and the use of sustainable manufacturing processes. Exploring bio-based coatings, eco-friendly surface treatments, and recycling methods for surface-engineered materials can contribute to a more sustainable approach.

Advancements in surface characterization techniques are crucial for understanding the structure–property relationships of surface engineered materials. Further development and refinement of characterization techniques, such as in situ and real-time monitoring methods, high-resolution imaging techniques, and advanced spectroscopic methods, will provide valuable insights into the behavior and performance of surface-engineered materials.

Exploring surface engineering solutions for extreme environments, such as high-temperature, corrosive, or abrasive conditions, is an area of great importance. Developing coatings and surface modifications that can withstand extreme conditions, maintain their functionality, and provide long-term protection is crucial for industries such as aerospace, energy, and manufacturing. Research efforts should focus on materials with high-temperature stability, superior corrosion resistance, and enhanced mechanical properties.

In the biomedical field, there is a need for advanced surface engineering techniques that can enhance the biocompatibility and bioactivity of materials. Developing novel coatings and surface modifications that promote cell adhesion, tissue integration, and controlled drug release is critical for applications such as implantable devices, tissue engineering, and regenerative medicine. Additionally, exploring surface engineering approaches to combat bacterial infections, such as developing antimicrobial surfaces and biofilm-resistant coatings, is of utmost importance.

Scale-Up and Industrial Implementation: For surface engineering techniques to have a significant impact in various industries, there is a need for scalable and easily implementable processes. Efforts should be directed towards developing surface engineering techniques that can be readily adopted by industrial manufacturers. This involves addressing challenges such as process scalability, cost-effectiveness, ease of integration into existing manufacturing processes, and robust quality-control measures.

The development of surfaces that can dynamically change their properties in response to external stimuli represents an exciting area for future research. This includes surfaces that can switch between different states of wettability, adhesion, or optical properties, enabling new functionalities and applications. Exploring materials with stimuli-responsive properties and developing surface engineering techniques to control their dynamic behavior will be crucial in this field.

Nature provides a wealth of inspiration for surface engineering. Future research should focus on understanding and replicating the unique properties and functionalities found in natural systems, such as self-cleaning, anti-fogging, or anti-reflective surfaces. By mimicking nature, researchers can develop innovative surface engineering solutions that offer improved performance, durability, and sustainability.

Many applications require surface engineering at multiple length scales, from the nanoscale to the macroscale. Research efforts should aim to develop techniques that can precisely control surface structures and properties across different length scales. This includes strategies for hierarchical surface engineering, where surface modifications are applied at different scales to achieve synergistic effects and optimized performance.

As additive manufacturing continues to advance, there is an opportunity to integrate surface engineering techniques directly into the 3D-printing process. This includes in situ surface modifications during printing or the development of novel printing materials with inherent surface functionalities. The combination of 3D printing and surface engineering can lead to the fabrication of complex components with tailored surface properties and functionalities. Overall, the future of surface engineering lies in addressing current limitations, exploring new application areas, and integrating emerging technologies. By focusing on sustainability, advanced characterization, artificial intelligence, extreme environments, biomedical applications, scalability, flexible electronics, dynamic surfaces, bio-inspired approaches, multiscale engineering, and 3D-printing integration, researchers can drive the field forward and unlock new possibilities for surface engineering in diverse industries and applications.

## 5. Conclusion

In conclusion, the exploration of interfacial engineering across various disciplines has unveiled a rich tapestry of strategies, challenges, and transformative potential. Key findings from the diverse realms of materials science, electronics, biomedical applications, and beyond underscore the central role that interfaces play in determining the performance, functionality, and adaptability of materials. Here, we summarize the key findings from our journey through interfacial engineering:

- Materials Science: Interfacial engineering in materials science is foundational to tailoring the properties of diverse materials, ranging from polymers to nanomaterials. Strategies such as surface modification, nanocomposite design, and nanoscale interactions enable precise control over material behaviors, unlocking enhanced mechanical, thermal, and optical properties.
- Electronics: Semiconductor interfaces and their optimization through interfacial engineering are pivotal for advancing electronic devices. Whether in the creation of efficient transistors, photovoltaic cells, or flexible electronics, engineered interfaces play a central role in determining device performance, reliability, and functionality.
- Biomedical Applications: In the biomedical arena, interfacial engineering emerges as a cornerstone for the development of biocompatible materials, drug delivery systems, and tissue engineering scaffolds. Surface modification strategies, the design of responsive interfaces, and the pursuit of biomimetic coatings showcase the potential to revolutionize healthcare by seamlessly integrating materials with biological systems.
- Challenges and Future Perspectives: While interfacial engineering holds immense promise, challenges such as the complexity of biological interfaces, long-term stability, and standardization persist. However, emerging trends and future directions, including the integration of nanotechnology, smart interfaces, artificial intelligence, and sustainable practices, point towards innovative solutions and the continued evolution of interfacial engineering.

The implications for future research in interfacial engineering are vast, offering a roadmap for exploration and innovation. As we look ahead, several key areas merit attention and investment:

- Multidisciplinary Collaboration: Interfacial engineering inherently demands collaboration across disciplines. Future research should encourage multidisciplinary teams, fostering synergy between materials scientists, biologists, engineers, and data scientists. This collaborative approach will accelerate the translation of research findings into practical applications.
- Integration of Emerging Technologies: The integration of emerging technologies, such as artificial intelligence, advanced characterization techniques, and sustainable practices, holds transformative potential. Future research should focus on harnessing these technologies to unravel the complexities of material interfaces and drive advancements in diverse fields.
- Biomimicry and Bio hybrid Designs: Exploration into biomimicry and bio hybrid designs offers a frontier for future research. Mimicking nature's design principles and integrating living and synthetic components present opportunities for creating materials with unparalleled functionalities and adaptability.
- Smart Materials and Responsive Interfaces: The development of smart materials with responsive interfaces represents a promising avenue. Future research should delve into the design of materials that can dynamically adapt to environmental stimuli, enabling precise control over functionalities in applications ranging from drug delivery to electronic devices.
- Sustainability and Green Practices: Sustainability is a growing concern in materials science. Future research should prioritize green interfacial engineering practices, exploring eco-friendly fabrication techniques, bio-based materials, and strategies for minimizing the environmental impact of interfacial engineering processes.
- Standardization and Reproducibility: Addressing challenges related to standardization and reproducibility is essential for the maturation of interfacial engineering. Future research should focus on establishing standardized protocols, refining fabrication techniques, and enhancing reproducibility to facilitate the seamless integration of interfacial engineering into various industries.

To sum up, the experience of interfacial engineering has given me a better understanding of the complex realm of material interfaces, how to manipulate them, and how they affect a wide range of applications. Interfacial engineering will play a key role in the advancement of science, technology, and societal well-being in the future, as researchers continue to solve the mysteries at the interfaces of materials and disciplines. The promise of novel materials, improved functionalities, and transformative technologies beckons.

## References


[1] A. Perez, A. Billard, C. Rébéré, C. Berziou, S. Touzain, and J. Creus, "Influence of metallurgical states on the corrosion behaviour of Al-Zn PVD coatings in saline solution," *Corros. Sci.*, 2013, doi: 10.1016/j.corsci.2013.04.048.

[2] R. Hausbrand, B. Bolado-Escudero, A. Dhont, and J. Wielant, "Corrosion of flame-assisted CVD silica-coated steel sheet," *Corros. Sci.*, vol. 61, pp. 28–34, 2012, doi: https://doi.org/10.1016/j.corsci.2012.04.007.

[3] G. A. Gaber, L. Z. Mohamed, A. Järvenpää, and A. Hamada, "Enhancement of corrosion protection of AISI 201 austenitic stainless steel in acidic chloride solutions by Ce-doped TiO2 coating," *Surf. Coatings Technol.*, 2021, doi: 10.1016/j.surfcoat.2021.127618.

[4] V. Stoilov and D. O. Northwood, "Friction, wear and corrosion: Learning from nature," *Int. J. Des. Nat. Ecodynamics*, 2014, doi: 10.2495/DNE-V9-N4-276-284.

[5] T. Liu, K. T. Lau, S. Chen, S. Cheng, and Y. Yin, "Super-hydrophobic surfaces improve corrosion resistance of Fe 3Al-type intermetallic in seawater," 2008. doi: 10.4028/www.scientific.net/amr.47-50.173.

[6] H. Herzallah, A. Elsayd, A. Shash, and M. Adly, "Effect of carbon nanotubes (CNTs) and silicon carbide (SiC) on mechanical properties of pure Al manufactured by powder metallurgy," *J. Mater. Res. Technol.*, 2020, doi: 10.1016/j.jmrt.2019.12.027.

[7] G. Fadel *et al.*, "Evaluation of corrosion and wear features of Al matrix reinforced with particles (SiC+Y2O3) coated with either nano-Ag/Ni or nano/Ag/Cu," *Egypt. J. Chem.*, vol. 67, no. 8, pp. 137–155, 2024, doi: 10.21608/ejchem.2024.247764.8845.

[8] B. Zaarour, L. Zhu, and X. Jin, "A Review on the Secondary Surface Morphology of Electrospun Nanofibers: Formation Mechanisms, Characterizations, and Applications," *ChemistrySelect*. 2020. doi: 10.1002/slct.201903981.

[9] B. Zaarour, L. Zhu, C. Huang, and X. Jin, "A mini review on the generation of crimped ultrathin fibers via electrospinning: Materials, strategies, and applications," *Polymers for Advanced Technologies*. 2020. doi: 10.1002/pat.4876.

[10] L. Z. Mohamed, G. Hamdy, and G. A. Gaber, "Performance of GO/SiO2, GO/TiO2, and GO/ZrO2 Nanocomposites Coatings as a Corrosion Barrier Layer on AlSi-Cu-Mg Alloy in 3.5 % NaCl Solution," *Int. J. Electrochem. Sci.*, vol. 16, pp. 1–18, 2021, doi: 10.20964/2021.05.13.

[11] G. A. Gaber, L. Z. Mohamed, and G. H. Abd El-Aziz Mohamed, "Enhancing Corrosion Behavior of T91 Steel in 3.5% NaCl through Graphene Oxide Nanocomposite Coatings," *Egypt. J. Chem.*, vol. 68, no. 2, pp. 75–89, 2025, doi: 10.21608/ejchem.2024.286273.9662.

[12] Y. Sasikumar, K. Indira, and N. Rajendran, "Surface Modification Methods for Titanium and Its Alloys and Their Corrosion Behavior in Biological Environment: A Review," *Journal of Bio- and Tribo-Corrosion*. 2019. doi: 10.1007/s40735-019-0229-5.

[13] B. Zaarour, L. Zhu, C. Huang, and X. Jin, "Fabrication of a polyvinylidene fluoride cactus-like nanofiber through one-step electrospinning," *RSC Adv.*, 2018, doi: 10.1039/c8ra09257e.

[14] B. Zaarour, L. Zhu, C. Huang, and X. Jin, "Controlling the Secondary Surface Morphology of Electrospun PVDF Nanofibers by Regulating the Solvent and Relative Humidity," *Nanoscale Res. Lett.*, 2018, doi: 10.1186/s11671-018-2705-0.

[15] L. Zhu, B. Zaarour, and X. Jin, "Direct generation of electrospun interconnected macroporous nanofibers using a water bath as a collector," *Mater. Res. Express*, 2020, doi: 10.1088/2053-1591/ab6c1e.

[16] B. Zaarour, L. Zhu, and X. Jin, "Direct fabrication of electrospun branched nanofibers with tiny diameters for oil absorption," *J. Dispers. Sci. Technol.*, 2021, doi: 10.1080/01932691.2020.1798779.

[17] B. Zaarour, "Enhanced piezoelectricity of PVDF nanofibers via a plasticizer treatment for energy harvesting," *Mater. Res. Express*, 2021, doi: 10.1088/2053-1591/ac3c72.

[18] H. M. El Refay, A. M. Hyba, and G. A. Gaber, "Fabrication, characterization and corrosion feature evaluation of mild steel in 1 M HCl by nanoparticle-modified activated carbon," *Chem. Pap.*, vol. 76, no. 2, pp. 813–825, 2022, doi: 10.1007/s11696-021-01895-2.

[19] G. A. Gaber, W. A. Hussein, and W. A. Ghanem, "Nanoparticles deposition effects on 1050A aluminum alloy anodic layers," *Mater. Res. Express*, vol. 7, no. 1, p. 16567, 2020, doi: 10.1088/2053-1591/ab6a59.

[20] B. Zaarour, L. Zhu, C. Huang, and X. Jin, "Enhanced piezoelectric properties of randomly oriented and aligned electrospun PVDF fibers by regulating the surface morphology," *J. Appl. Polym. Sci.*, 2019, doi: 10.1002/app.47049.

[21] B. Zaarour and W. Liu, "Enhanced piezoelectric performance of electrospun PVDF nanofibers by regulating the solvent systems," *J. Eng. Fiber. Fabr.*, 2022, doi: 10.1177/15589250221125437.

[22] B. Zaarour, L. Zhu, and X. Jin, "Controlling the surface structure, mechanical properties, crystallinity, and piezoelectric properties of electrospun PVDF nanofibers by maneuvering molecular weight," *Soft Mater.*, 2019, doi: 10.1080/1539445X.2019.1582542.

[23] B. Zaarour, L. Zhu, and X. Jin, "Direct generation of electrospun branched nanofibers for energy harvesting," *Polym. Adv. Technol.*, 2020, doi: 10.1002/pat.4992.

[24] B. Zaarour and W. Liu, "Recent advances of textile sorbents for oil spills cleanup: A review," *Journal of Industrial Textiles*. 2023. doi: 10.1177/15280837231186652.

[25] B. Zaarour, H. Tina, L. Zhu, and X. Y. Jin, "Branched nanofibers with tiny diameters for air filtration via one-step electrospinning," *J. Ind. Text.*, 2022, doi: 10.1177/1528083720923773.

[26] L. Zhu, B. Zaarour, and X. Jin, "Fabrication of perfect CMCS/PVA nanofibers for keeping food fresh via an in situ mixing electrospinning," *Mater. Res. Express*, 2019, doi: 10.1088/2053-1591/ab5396.

[27] J. H. Park *et al.*, "Defect passivation of transition metal dichalcogenides via

a charge transfer van der Waals interface," *Sci. Adv.*, 2017, doi: 10.1126/sciadv.1701661.

[28] Y. Yang, J. Li, L. Lin, and X. Peng, "An efficient and surface-benign purification scheme for colloidal nanocrystals based on quantitative assessment," *Nano Res.*, 2015, doi: 10.1007/s12274-015-0835-6.

[29] C. Soci *et al.*, "ZnO nanowire UV photodetectors with high internal gain," *Nano Lett.*, 2007, doi: 10.1021/nl070111x.

[30] F. H. Alsultany, Z. Hassan, and N. M. Ahmed, "A high-sensitivity, fast-response, rapid-recovery UV photodetector fabricated based on catalyst-free growth of ZnO nanowire networks on glass substrate," *Opt. Mater. (Amst).*, 2016, doi: 10.1016/j.optmat.2016.07.004.

[31] L. M. Shaker, A. Al-Amiery, M. S. Takriff, W. N. R. W. Isahak, and W. K. Al-Azzawi, "Clear vision, green choices: a review of optics, environmental effects, and eco-friendly advancements in contact lenses," *J. Opt.*, 2023, doi: 10.1007/s12596-023-01290-y.

[32] A. S. Mahdi, L. M. Shaker, and A. Alamiery, "Recent advances in organic solar cells: materials, design, and performance," *J. Opt.*, 2024, doi: 10.1007/s12596-023-01262-2.

[33] G. K. Shamnamol, K. P. Sreelakshmi, G. Ajith, and J. M. Jacob, "Effective utilization of drugs as green corrosion inhibitor-A review," 2020. doi: 10.1063/5.0005931.

[34] G. A. Gaber, M. M. Soliman, Z. A. Nasr, and A. M. Hyba, "Comprehensive investigation of sustainable corrosion inhibitors on Cu–Zn alloy in simulated cooling water: Electrochemical explorations, SEM/EDX analysis, and DFT/molecular simulations utilizing expired Bepotastine-B as a green inhibitor," *Sustain. Chem. Pharm.*, 2024, doi: 10.1016/j.scp.2023.101340.

[35] Z. Hu, Z. Wu, C. Han, J. He, Z. Ni, and W. Chen, "Two-dimensional transition metal dichalcogenides: Interface and defect engineering," *Chemical Society Reviews*. 2018. doi: 10.1039/c8cs00024g.

[36] H. M. Abd El-Lateef, K. Shalabi, A. R. Sayed, S. M. Gomha, and E. M.

Bakir, "The novel polythiadiazole polymer and its composite with α-Al(OH)3 as inhibitors for steel alloy corrosion in molar H2SO4: Experimental and computational evaluations," *J. Ind. Eng. Chem.*, 2022, doi: 10.1016/j.jiec.2021.09.022.

[37] M. Madani *et al.*, "Green synthesis of nanoparticles for varied applications: Green renewable resources and energy-efficient synthetic routes," *Nanotechnology Reviews*. 2022. doi: 10.1515/ntrev-2022-0034.

[38] H. M. Dardeer, S. A. Abbas, M. M. Ghobashy, G. A. Gaber, and M. F. Aly, "Synthesis and characterization of novel chitosan-sodium pyruvate polymer and its derivatives for corrosion feature evaluation of Cu-Ni alloy," *Inorg. Chem. Commun.*, 2023, doi: 10.1016/j.inoche.2023.111308.

[39] G. A. Gaber, W. A. Hussein, and A. S. I. Ahmed, "Comparative study of electrochemical Corrosion of Novel Designs of 90/10 and 70/30 Copper-Nickel Alloys in Brine Solutions," *Egypt. J. Chem.*, vol. 63, no. 4, pp. 1527–1540, 2020, doi: 10.21608/ejchem.2019.16389.2001.

[40] D. Zhang *et al.*, "Ultraviolet photodetection properties of indium oxide nanowires," *Appl. Phys. A Mater. Sci. Process.*, 2003, doi: 10.1007/s00339-003-2099-3.

[41] D. Barettin *et al.*, "Carrier transport and emission efficiency in InGaN quantum-dot based light-emitting diodes," *Nanotechnology*, 2017, doi: 10.1088/1361-6528/aa75a8.

[42] X. Yang *et al.*, "Efficient green light-emitting diodes based on quasi-two-dimensional composition and phase engineered perovskite with surface passivation," *Nat. Commun.*, 2018, doi: 10.1038/s41467-018-02978-7.

[43] X. Yang *et al.*, "Erratum: Author Correction: Efficient green light-emitting diodes based on quasi-two-dimensional composition and phase engineered perovskite with surface passivation (Nature communications (2018) 9 1 (570))," *Nature communications*. 2018. doi: 10.1038/s41467-018-03702-1.

[44] L. M. Shaker, A. Alamiery, M. S. Takriff, and W. N. R. W. Isahak, "Integrated TiO2 nanoparticle for electronic polymeric contact lenses," *J. Opt.*, 2023, doi: 10.1007/s12596-023-01195-w.

[45] Y. Jin *et al.*, "Study of Carbon Nanotubes as Etching Masks and Related Applications in the Surface Modification of GaAs-based Light-Emitting Diodes," *Small*, 2015, doi: 10.1002/smll.201500869.

[46] J. T. Lim *et al.*, "Unraveled Face-Dependent Effects of Multilayered Graphene Embedded in Transparent Organic Light-Emitting Diodes," *ACS Appl. Mater. Interfaces*, 2017, doi: 10.1021/acsami.7b13973.

[47] B. R. Lee *et al.*, "Amine-based interfacial molecules for inverted polymer-based optoelectronic devices," *Adv. Mater.*, 2015, doi: 10.1002/adma.201500663.

[48] X. Zhang *et al.*, "Bright Perovskite Nanocrystal Films for Efficient Light-Emitting Devices," *J. Phys. Chem. Lett.*, 2016, doi: 10.1021/acs.jpclett.6b02073.

[49] J. S. Yao *et al.*, "Suppressing Auger Recombination in Cesium Lead Bromide Perovskite Nanocrystal Film for Bright Light-Emitting Diodes," *J. Phys. Chem. Lett.*, 2020, doi: 10.1021/acs.jpclett.0c02777.

[50] Z. Chen *et al.*, "High-Performance Color-Tunable Perovskite Light Emitting Devices through Structural Modulation from Bulk to Layered Film," *Adv. Mater.*, 2017, doi: 10.1002/adma.201603157.

[51] L. M. Shaker, A. A. Al-Amiery, and W. K. Al-Azzawi, "A clearer vision: a mini-review on contact lenses," *J. Opt.*, 2024, doi: 10.1007/s12596-023-01222-w.

[52] L. M. Shaker, A. Al-Amiery, W. N. R. W. Isahak, and W. K. Al-Azzawi, "Advancements in Quantum Optics: Harnessing the Power of Photons for Next-Generation Technologies," *J. Opt.*, 2023, doi: 10.1007/s12596-023-01320-9.

[53] M. S. Idris, S. Subramani, W. M. W. A. Kamil, and M. Devarajan, "Synthesis of MgO thin film on aluminum and copper substrates as thermal interface materials," *IEEE Trans. Electron Devices*, 2019, doi: 10.1109/TED.2019.2895622.

[54] F. N. Alaribe, S. L. Manoto, and S. C. K. M. Motaung, "Scaffolds from biomaterials: Advantages and limitations in bone and tissue engineering,"

*Biologia (Bratisl).*, 2016, doi: 10.1515/biolog-2016-0056.

[55] B. M. Holzapfel *et al.*, "How smart do biomaterials need to be? A translational science and clinical point of view," *Advanced Drug Delivery Reviews*. 2013. doi: 10.1016/j.addr.2012.07.009.

[56] S. Bose, S. F. Robertson, and A. Bandyopadhyay, "Surface modification of biomaterials and biomedical devices using additive manufacturing," *Acta Biomaterialia*. 2018. doi: 10.1016/j.actbio.2017.11.003.

[57] P. Roach, D. Eglin, K. Rohde, and C. C. Perry, "Modern biomaterials: A review - Bulk properties and implications of surface modifications," *Journal of Materials Science: Materials in Medicine*. 2007. doi: 10.1007/s10856-006-0064-3.

[58] G. Wu, P. Li, H. Feng, X. Zhang, and P. K. Chu, "Engineering and functionalization of biomaterials via surface modification," *Journal of Materials Chemistry B*. 2015. doi: 10.1039/c4tb01934b.

[59] D. Katti, R. Vasita, and K. Shanmugam, "Improved Biomaterials for Tissue Engineering Applications: Surface Modification of Polymers," *Curr. Top. Med. Chem.*, 2008, doi: 10.2174/156802608783790893.

[60] L. M. Shaker, A. Al-Amiery, W. N. R. W. Isahak, and W. K. Al-Azzawi, "Integrated photonics: bridging the gap between optics and electronics for enhancing information processing," *J. Opt.*, 2023, doi: 10.1007/s12596-023-01321-8.

[61] L. M. Shaker, A. H. Al-Hamdani, and A. A. Al-Amiery, "A comparative study of optical quality for different polymeric contact lenses," 2019. doi: 10.1088/1742-6596/1234/1/012042.

[62] L. M. Shaker, A. Alamiery, M. Takriff, and W. N. R. Wan Isahak, "Novel blue-wavelength-blocking contact lens with er3+/tio2 nps: Manufacture and characterization," *Nanomaterials*, 2021, doi: 10.3390/nano11092190.

[63] N. Huang, Y. Wang, Y. Zhang, L. Liu, N. Yuan, and J. Ding, "Multifunctional coating on magnesium alloy: Superhydrophobic, self-healing, anti-corrosion and wear-resistant," *Surf. Coatings Technol.*, 2023, doi: 10.1016/j.surfcoat.2023.129539.

[64] X. Sun, M. Jing, H. Dong, W. Xie, and F. Luo, "CuO-ZnO submicroflakes with nanolayered Al2O3 coatings as high performance anode materials in lithium-ion batteries," *J. Alloys Compd.*, 2023, doi: 10.1016/j.jallcom.2023.170137.

[65] D. Sun and K. F. Böhringer, "Self-cleaning: From bio-inspired surface modification to MEMS/microfluidics system integration," *Micromachines*. 2019. doi: 10.3390/mi10020101.

[66] M. Ramezani and Z. Mohd Ripin, "4D Printing in Biomedical Engineering: Advancements, Challenges, and Future Directions," *Journal of Functional Biomaterials*. 2023. doi: 10.3390/jfb14070347.

[67] B. Mutharani, P. Ranganathan, T. W. Chen, S. M. Chen, and J. Vinoth Kumar, "Thermoreversible Switchlike Electrocatalytic Reduction of Tizanidine Based on a Graphene Oxide Tethered Stimuli-Responsive Smart Surface Supported Pd Catalyst," *Anal. Chem.*, 2020, doi: 10.1021/acs.analchem.0c00958.

[68] H. Li *et al.*, "Machine learning assisted predicting and engineering specific surface area and total pore volume of biochar," *Bioresour. Technol.*, 2023, doi: 10.1016/j.biortech.2022.128417.

FSC
www.fsc.org
MIX
Papier | Fördert gute Waldnutzung
FSC® C083411

Kolibri 360 GmbH
Ferdinand-Jühlke-Straße 7
99095 Erfurt, Deutschland
produktsicherheit@kolibri360.de

Druck:
CPI Druckdienstleistungen GmbH
im Auftrag der
Kolibri 360 GmbH
Ferdinand-Jühlke-Str. 7
99095 Erfurt